American Wood Working Machine Co.

New and Improved Wood Working Machinery

Illustrated Catalogue

American Wood Working Machine Co.

New and Improved Wood Working Machinery
Illustrated Catalogue

ISBN/EAN: 9783741105593

Manufactured in Europe, USA, Canada, Australia, Japa

Cover: Foto ©Andreas Hilbeck / pixelio.de

Manufactured and distributed by brebook publishing software
(www.brebook.com)

American Wood Working Machine Co.

New and Improved Wood Working Machinery

American Wood Working Machine Co.

———◦ SUCCESSOR TO ◦———

F. H. CLEMENT CO.,	ROWLEY & HERMANCE CO.,	MILWAUKEE SANDER MFG. CO.,
GLEN COVE MACHINE CO.,	HOYT & BRO. CO.,	C. B. ROGERS & CO.,
GOODELL & WATERS,	LEHMAN MACHINE CO.,	THE LEVI HOUSTON CO.,
YOUNG BROS. CO.,	WILLIAMSPORT MACHINE CO.	

ILLUSTRATED CATALOGUE

——— OF ———

NEW AND IMPROVED

WOOD WORKING MACHINERY.

OFFICES:

Washington Life Building, corner Broadway and Liberty Street,

NEW YORK CITY.

SALESROOMS:

109 Liberty Street, New York City.   43 and 45 S. Canal Street, Chicago, Ill.

94 Pearl Street, Boston, Mass.   Church and Basin Streets, Williamsport, Pa.

Cable Address: WOODMACHO.   FIRST EDITION, 1898.

❧ INTRODUCTORY. ❧

To Our Patrons:

The American Wood Working Machine Company is composed of the following well known firms

FRANK H. CLEMENT CO.	ROWLEY & HERMANCE CO.	MILWAUKEE SANDER MFG. CO.
GLEN COVE MACHINE CO.	HOYT & BRO. CO.	C. B. ROGERS & CO.
GOODELL & WATERS.	LEHMAN MACHINE CO.	THE LEVI HOUSTON CO.
YOUNG BROS. CO.		WILLIAMSPORT MACHINE CO.

The lines of Machines found in the ensuing pages have been selected from the above named plants on account of their superiority over all other machines of their kind, as to their mechanical perfection and power to perform the work for which they are built.

This Company is not a Trust, and does not intend to make its money by inflating prices; all we want is a fair profit, and in return we give our customers the benefit of our consolidation, namely: economical manufacture, consolidated experience and a line so large to select from that our salesmen are able to give you an unbiased opinion as to the best machine or machines for your use.

We have not by any means established a standard, for our aim is to improve our machines to the extent that money and brains will bring them.

We have established Salesrooms at 109 Liberty Street, New York; 43 and 45 South Canal Street, Chicago, Ill.; 94 Pearl Street, Boston, Mass., and Church and Basin Streets, Williamsport, Pa. All sales communications should be addressed to our salesroom nearest to you, thus facilitating prompt attention and immediate interview by one of our salesmen when so desired.

Owing to the demand for a catalogue we have been obliged to compile this, our first issue, in a hurried manner, and desire to say that it does not represent our full line of machines. As soon as convenient we shall issue a complete catalogue comprising all of the various machines made by our separate Branches.

Thanking you for past favors and trusting for a continuance of your patronage, we remain

Very truly yours,

AMERICAN WOOD WORKING MACHINE CO.

4

PLEASE BE KIND ENOUGH to acknowledge receipt of this Catalogue by postal card or letter. If in want of any machine herein illustrated, please mention it, and we will quote you prices.

Suggestions to Correspondents.

To avoid mistakes, give your post-office address in full; TOWN, COUNTY AND STATE.

In ordering extras, repairs, supplies, or changes in machines being built, give sufficient information to enable us to fill the order intelligently and correctly.

When sending anything to us, put your name and address on each package.

Guess Work is Expensive.

Orders thoughtlessly or carelessly made are often received with instructions to hurry the goods forward. To avoid delay we must GUESS what is wanted, and we may guess wrong; then comes the inevitable expense of express charges, delay and consequent vexations at both ends of the line.

We will send EXACTLY what you want if you will state your wants EXACTLY.

In ordering gears give number of teeth, diameter, width of face, size of hole and length of hub. If a pulley, give diameter, face and hole, state whether straight or crown face; and if it is to run with another pulley (tight or loose), give projection of hub beyond the rim. If saws, always state whether rip or cross-cut, gauge, diameter of hole and number of teeth. In ordering mortise chisels, say whose make of mortiser they are for, or send an old chisel shank so we can get the proper taper. Orders for moulding bits should be accompanied with patterns, showing position and size of slots, as the spring of the bit depends upon the diameter of the cutting circle of the head. When ordering blanks for moulding bits, always give the length, as we do not know the thickness of moulding for which they will be shaped.

In ordering knives or moulding bits for our make of machines, state which machine they are for. If for other makes of machines, give diameter of head, and state whether the head is solid or slotted, the size, number and position of the slots. If possible, always send a paper pattern of the full size knife that can be used.

A LITTLE PENCIL SKETCH often conveys more meaning than a page of written instructions.

Many of our machines have been sold through agents, and only the agents' names are known to us in the transaction, hence the necessity of the greatest care in ordering.

A CAREFUL CONSIDERATION of the foregoing will save time and correspondence, and enable us to fill orders promptly.

Terms of Sale.

Strangers in ordering will facilitate shipment by sending reference.

All bills are due and subject to sight draft thirty days from date of shipment, unless other terms are agreed upon. Thirty days' time is extended to purchasers favorably rated in Commercial Reports.

CLAIMS FOR ERRORS must be made on receipt of goods. Our responsibility ceases upon delivery of shipment in good order to the Transportation Company, and in no case can we allow claims for damages or loss in transit. All claims *must* be made to the Transportation Company.

If in want of any kind of machinery for working wood, write us for prices before placing order. We will save you money, and give you the best the market affords.

Respectfully yours,

AMERICAN WOOD WORKING MACHINE COMPANY.

5

Cable Address, "Woodmacho." | **GENERAL CODE.** | A. B. C. Code Used Also.

CODE TABLE OF DATES.

CODE WORD.	DAY.	CODE WORD.	DAY.
Dubbing	1st.	Daintre	17th.
Dabbler	2d.	Dairy	18th.
Dabchick	3d.	Dairymaid	19th.
Dabovis	4th.	Dairyman	20th.
Dactillon	5th.	Daisied	21st.
Dadais	6th.	Dallage	22d.
Daddy	7th.	Dallador	23d.
Daft	8th.	Dalliance	24th.
Dagger	9th.	Dallying	25th.
Daglock	10th.	Damask	26th.
Dagorne	11th.	Dame	27th.
Dahlia	12th.	Dameret	28th.
Daigner	13th.	Damper	29th.
Dainties	14th.	Dampness	30th.
Daintily	15th.	Damsel	31st.
Daintiness	16th.		

CODE WORD.	MONTH.	CODE WORD.	MONTH.
Dancers	January.	Dangerless	July.
Dandelion	February.	Dangle	August.
Dandified	March.	Dansant	September.
Dandler	April.	Dapper	October.
Dandruff	May.	Darder	November.
Danewort	June.	Dareful	December.

SHIPMENTS.

CODE WORD.		CODE WORD.	
Darling	Wire earliest possible shipment of	Debark	Ship in carload.
Daringness	How soon can you ship?	Debarring	When and by what route did you ship?
Darken	We can ship at once.	Debased	Will a few days' delay in shipment make any difference?
Darkling	Expect to ship	Debasement	A few days' delay in shipment will make no
Darkness	Will ship		difference.
Darksome	Can ship	Debatable	We have nearly finished and can ship
Darn	Can ship in three days.	Debating	None in stock, but can ship in
Darning	Can ship in five days.	Debel	Reply by telegraph when you can ship.
Dashing	Can ship in one week.	Debilement	When will you ship?
Dateless	Can ship in ten days.	Debility	We ship to-day
Dative	Can ship in two weeks.	Debitage	Goods were shipped.
Daubing	Can ship in three weeks.	Debris	If you can ship at once, do so.
Daughter	Can ship in four weeks.	Debtor	We have shipped your order of
Dauntless	Can ship in five weeks.	Debutant	We have not shipped.
Dauphin	Can ship in six weeks.	Decadency	If you ship
Dawning	Can ship in two months.	Decalogue	Ship and draw with bill of lading attached.
Daybook	Can ship in three months.	Decameron	Can you ship and get through bill of lading?
Daybreak	Ship by steamer from	Decamp	Will ship as soon as possible.
Daydream	Ship by sailing vessel from	Decampment	Prepare to ship, but wait further instructions
Daylight	Ship by steamer to		by mail.
Daytime	Ship by sailing vessel to	Decanter	If you have not shipped, await further advice.
Dazzle	Ship by express at once.	Decapage	goods are ready to ship and we await
Dazzling	Ship by boat at once.		further instructions
Deacon	Ship by express at once.	Decapitate	Shall we ship?
Deadish	Ship by express C. O. D.	Decayed	Hurry shipment as much as possible.
Deadman	Shall we ship by steamer or sailing vessel?	Deceitful	If you have not shipped
Deafen	Shall we ship by rail or boat?	Deceiver	Will be shipped this week.
Dealer	Ship by sailing vessel from New York.	Deceiving	Can ship at once.
Dean	Ship by steamer from New York.	Decency	Will be shipped next week.
Deanery	Ship by cheapest route.	Decently	When ready to ship, telegraph us.
Deanship	Ship by railroad if not too great a difference	Deception	Shipment can be made by
	in cost.	Deceptive	Send shipping directions.
Dearling	Ship by fast freight.	Decevant	Shipment delayed by
Dearness	Ship by steamer and insure.	Decimal	Car load.
Deathless	Ship by sailing vessel and insure.	Decharne	In one shipment.
Deathlike	Ship as soon as possible.	Dechirage	Do not ship.
Deathly	Ship immediately.	Decidence	If you have not shipped will send
Deathwatch	Ship immediately, without waiting for car-		from here, answer.
	load.		

INSURANCE.

CODE WORD.		CODE WORD.	
Deedless	Shall we insure?	Deepness	You may insure.
Deemster	Do not insure.	Deerfold	Insure and let charges follow.

AMERICAN WOOD-WORKING MACHINE CO.

GENERAL CODE—Continued.

A. B. C. Code Used Also.

FREIGHT RATES, WEIGHTS, MEASUREMENTS AND HORSE POWER.

CODE WORD.

Defence............ What is the best rate of freight you can obtain from your place to ....?
Defendable...... Rate of freight per 100 pounds, in carload lots, from .... to .... is ....
Defendant...... Rate of freight per 100 pounds, in carload lots released from .... to .... is ....
Defensive........ Rate of freight per 100 pounds, in less than carload lots, from .... to .... is ....
Defensory........ Rate of freight per 100 pounds, in less than carload lots released, from .... to .... is ....

CODE WORD

Defile............ What is cubic feet and weight of.........?
Defilement.........Cubic feet ....
Definable........ What is shipping weight of ....?
Definative........ Shipping weight is ....?
Definite........ What is approximate gross weight?
Defleet............ What is approximate gross weight of ....?
Deflecting........ What is net weight of........?
Deflexion........What is net and gross weight of ....?
Deflower........What horse power is required for ....?
Defluous........ horse power is required to drive to full capacity.

PRICES.

CODE WORD.

Deftly............ What is the lowest net cash price for ....?
Defunct............What is the lowest net cash price delivered here on ....?
Defying............ The lowest net cash price on .... is ....
Degenerate........ Wire lowest net cash price to us delivered here on ....
Degluer......... Is price net, or list subject to discount?
Degollar......... Price quoted you is net to us F. O. B. cars at........

CODE WORD.

Degommage........ Price quoted you is net to us F. O. B. cars at New York City.
Degonfler........ Price quoted is net cash, boxed for export and delivered on cars at New York City.
Degout............Telegraph price on ....
Degoutant........Telegraph lowest price on ....
Degradant........ Your price is too high.

SALES.

CODE WORD.

Deist............ Sell at price that will net us not less than ....
Deistical........ Do not sell unless you can get our price.
Dejected........ Sell at price you name if you cannot do better.
Dejeuner........ You may sell at........
Delantera........ Sell at your discretion.
Delarder........ Do not sell for less than........
Delate........ Cannot sell at........
Delayant........ Cannot sell at price you name, but can get........
Delayement........Cannot sell at over........

CODE WORD.

Deleatur........ Have you sold?
Delisser........ Have sold.
Delectable.. Have not sold.
Delegated,. Have sold and wish you to replace ....
Delestage..... If you have not sold,
Deletory........ All in the works are sold.
Delft............ Will accept. Have written.
Delfinio........ Will not accept. Have written.
Delgado........Cannot accept. Have written.

PAYMENTS, REFERENCES, ETC.

CODE WORD.

Demater............Have you remitted by mail?
Demeaning........Have remitted by mail.
Demeanour........Will remit at once.
Demediar........Draw at sight.
Demenage........Draw at sight Bill of Lading attached.
Demency........Have drawn at three days' sight Bill of Lading attached.
Demented........Draw on us at........
Demerit........ Have drawn on you at sight.
Demeubler........Have drawn on you as usual.
Demise........Have drawn on you at sight with Bill of Lading attached.
Demisable........How shall we draw?
Demisory........Have remitted through ....
Democracy........Shall we draw at sight?
Democratic........Will draw on you as per proposition.
Demolish........Shall we draw on you?
Demolition........How much shall we draw for?
Demonian........Have you drawn?
Demonstrat........How have you drawn?
Demoralize........How much did you draw for?

CODE WORD.

Dempster........Have you remitted?
Demurely........Will you favor us with check in settlement of invoice........?
Demureness........Will you favor us with check in settlement of balance?
Deniable........Remit without delay.
Denial........Have placed credit for........dollars with........
Denigrate........You may open credit in our favor with New York Bank for........
Denization........Have opened Bank credit in your favor with........
Denizen........Have mailed you Bankers' draft on New York for........, draw at sight against Bill of Lading for balance.
Denominate........Have mailed you Bankers' draft on London for........, draw at sight against Bill of Lading for balance.
Denotable...... We refer to...........
Denoter........We refer to...........New York.
Denouncer........Buy for me (or on our account) and hold for shipping instructions, as follows.....

7

Fig. 1.

F. H. CLEMENT CO.'S

New 54-Inch Band Re=Sawing Machine.

THE PERFECTION OF DESIGN AND WORKMANSHIP.

THIS machine will be found on examination to be a long step in advance of any tool of its kind yet produced in the way of simplicity and directness of action, and elegance and adaptability of design. It embodies all the conveniences and attachments that are necessary or desirable for any kind of resawing in hard or soft wood, and it has ample power both on the blade and in the feed works for any reasonable demand; at the same time the cost has been kept below that of other machines that have far less merit.

The **Frame** is cast hollow with cross struts and heavy foot flanges, and it has a broad base which when properly set does not permit vibration of the machine when running.

The **Shafts** are of hammered crucible steel, and the lower one is 3½ inches diameter, having three bearings, each 12 inches long, with automatic oiling cells and return channels. The main upper bearing is also 12 inches in length with similar self-oiling attachments.

The **Wheels** are of a form and dimensions which have been found correct in experience, and they are both "dished" so as to extend over the boxes, thereby bringing the strain of the blade directly on the bearings. The lower wheel is very heavy, with a solid central web, and the upper one is as light as possible consistent with strength.

The **Feed Works** are driven by friction gearing, which is adjustable to vary the feed from 18 feet to 100 lineal feet per minute, and the arrangement is the most simple possible, every adjustable part being within easy reach of the operator at his post. The rolls are driven by bronze spur gears and steel worms with ball end-bearings, and the motion is perfectly smooth and noiseless even at the fastest speed.

Six Feed Rolls carry the stock to the saw, four of them being 5 inches diameter and driven by gearing, and two of them solid steel idle guide-rolls close to the saw. The right hand set of rolls are rigid in their boxes, but the left hand set are elastic so as to grasp uneven stock and hold it fairly up against the rigid rolls, thus making a powerful feed even on very unequally sawed lumber. The rolls tilt to an angle to saw clapboards and the forward pair can be fluted if so ordered.

The **Self-Centering Attachment** is so arranged that by slacking a set screw and adjusting a collar, the right hand rolls become rigid, but may be adjusted to thickness by the lower screw and hand crank.

The **Straining of the Blade** is accomplished by a balance lever with weights which may be changed according to the work and width of the saw. There is also a cushioned connection between the tilting screw and upper box, permitting sufficient elasticity for the protection of the blade.

The **Guides** have large hardened steel rear or safety rollers and independent side guides which are adjustable by screws. The lower guide forms a work table for the lumber passing through, and the upper one adjusts to receive stock 30 inches wide and is counter-balanced and adjusted vertically by a large pilot wheel or by a lever and bar overhead as desired; thus the guide can be instantaneously shifted as the lumber varies in width.

The **Capacity** of the machine is 30 inches wide (or deep) and to the center of 16 inches between the rolls, or stock 8 inches thick can be "slabbed off" ⅞ inch and thicker, by setting the right hand rolls rigid. Dry pine or similar soft woods 10 inches wide or less has been sawed on this machine at 75 to 85 lineal feet per minute, or at the rate of 48,000 surface feet per day; and the same kind of stock 12 to 16 inches wide has been cut at 60 to 70 feet per minute, or at the rate of 52,000 surface feet per day. For kiln dried hard woods the speed of feed will necessarily be much less.

Blades 5 inches wide and under can be used, 18 to 22 gauge thick. We recommend blades 5 inches wide and No. 19 gauge, and we send one blade complete, ready for sawing, with each machine.

The **Driving Pulleys** are 24x8½ inches, and the loose pulley is 1 inch smaller than the tight and has a self-oiling detachable bush, lined with genuine babbitt. They should run 500 to 550 per minute. The machines are shipped crated and taken apart as far as possible, and we furnish ample directions for setting and operating with necessary floor plans. The engraving shows only the driving pulley, designed to be used with a tightener, but we furnish loose pulley without extra charge.

The **Workmanship** is superior to that of any other machine of its kind now made, and we shall be glad to show any prospective buyer our methods.

☞ In place of the pilot wheel and shaft for adjusting the upper guides, we can furnish when wanted a T lever attachment to be hung from the ceiling, which is used with a horizontal shifting bar extending back within reach of the operator.

	Weight.	Code Word.
Fig. 1.—Complete, 54-inch Band Re-Sawing Machine	7,200 lbs.	**Fabago.**

13

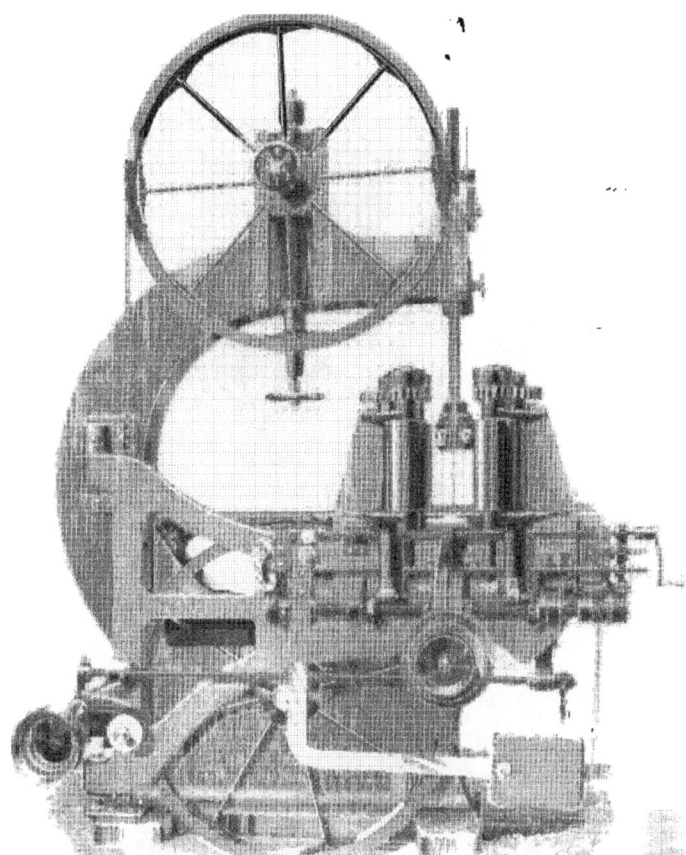

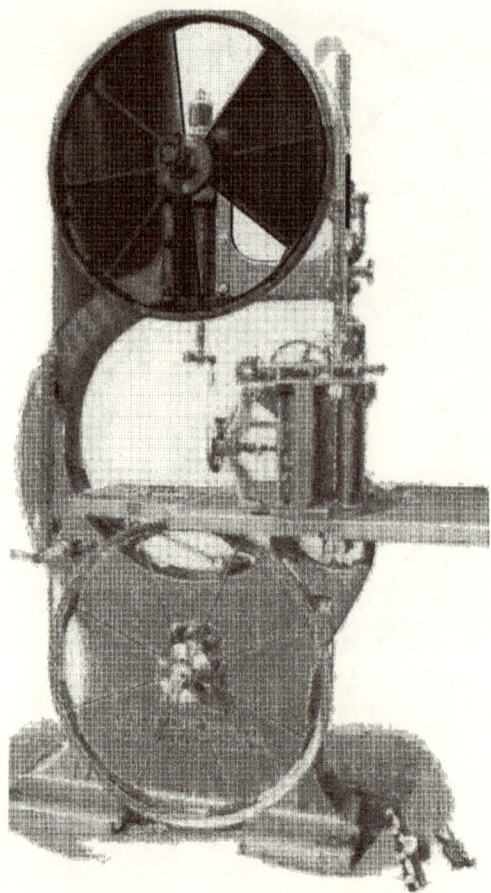

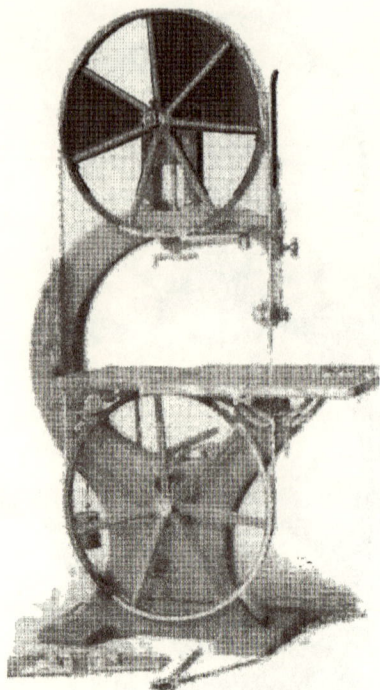

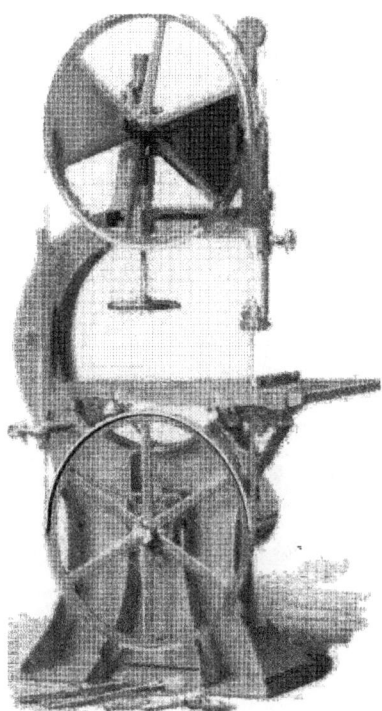

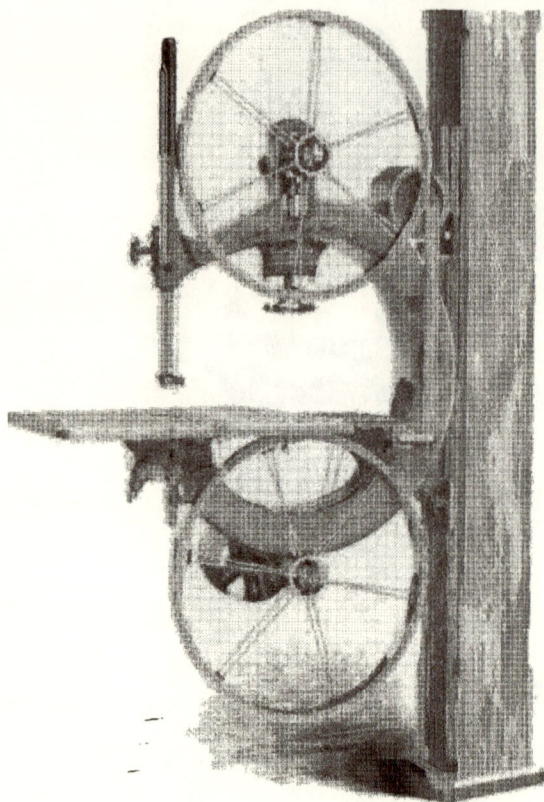

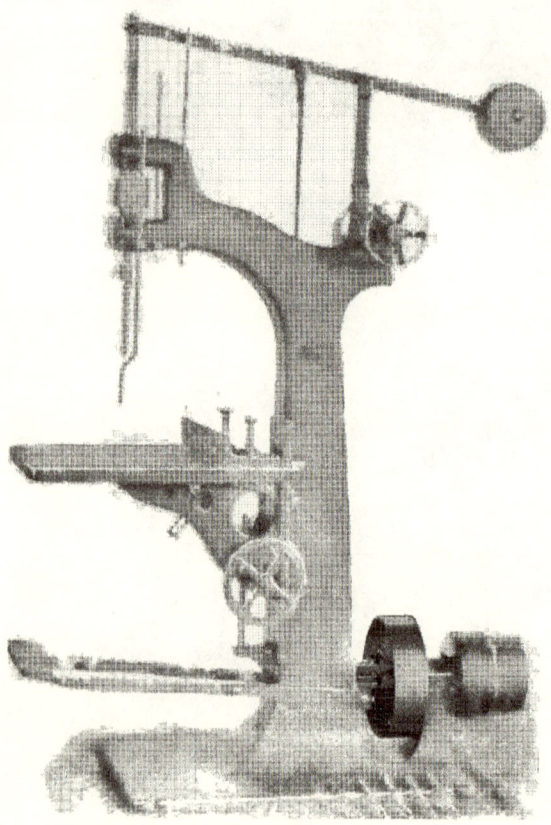

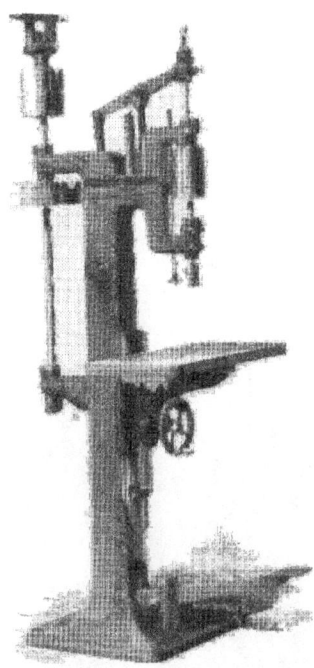

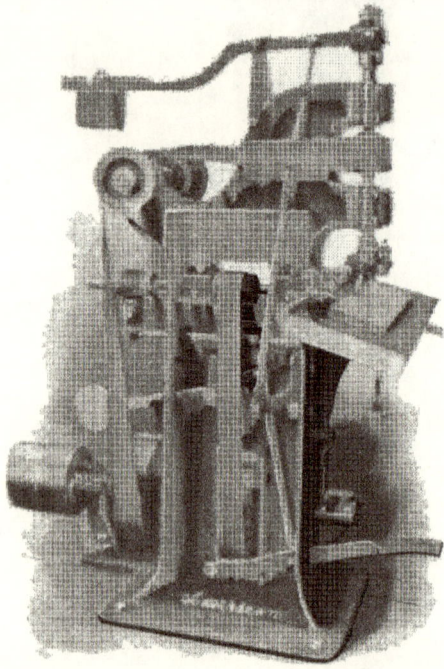

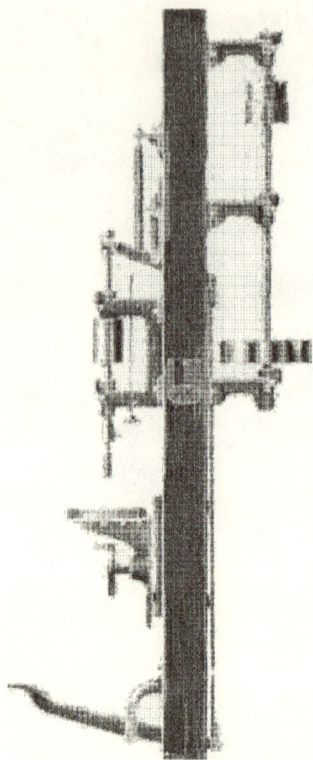

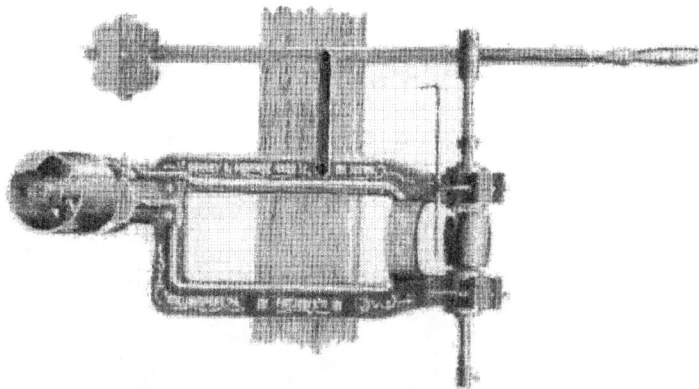

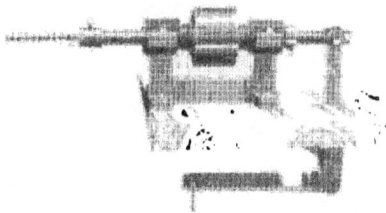

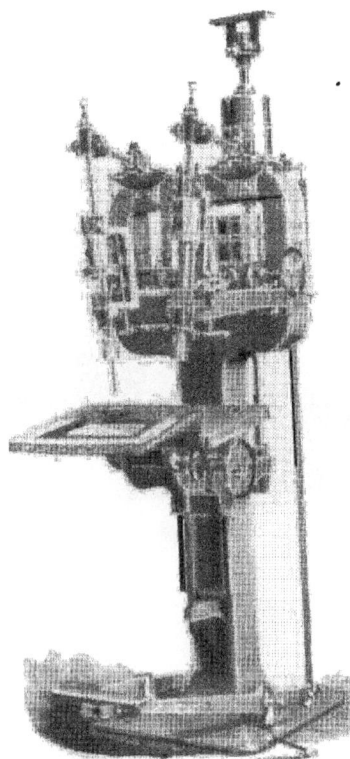

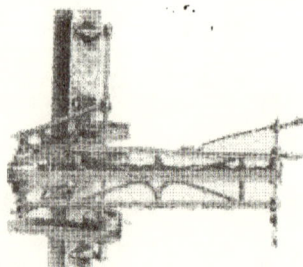

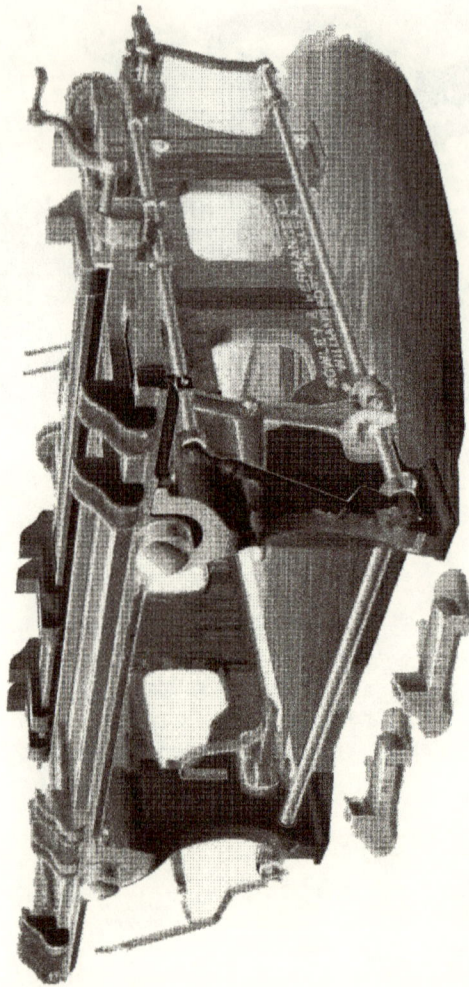

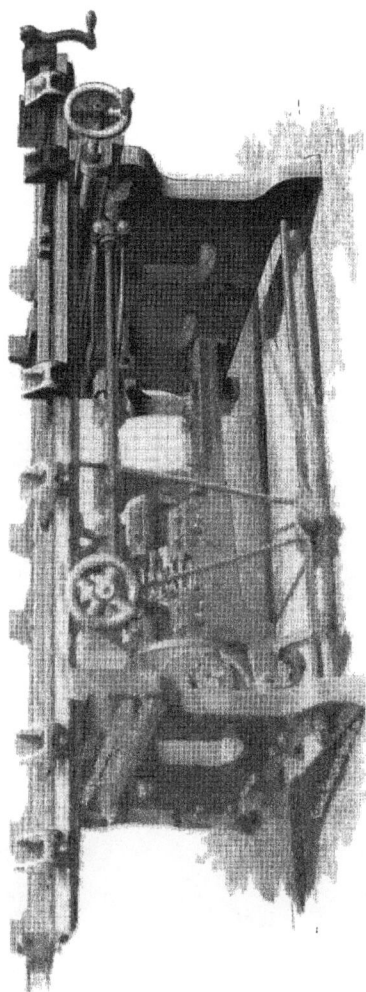

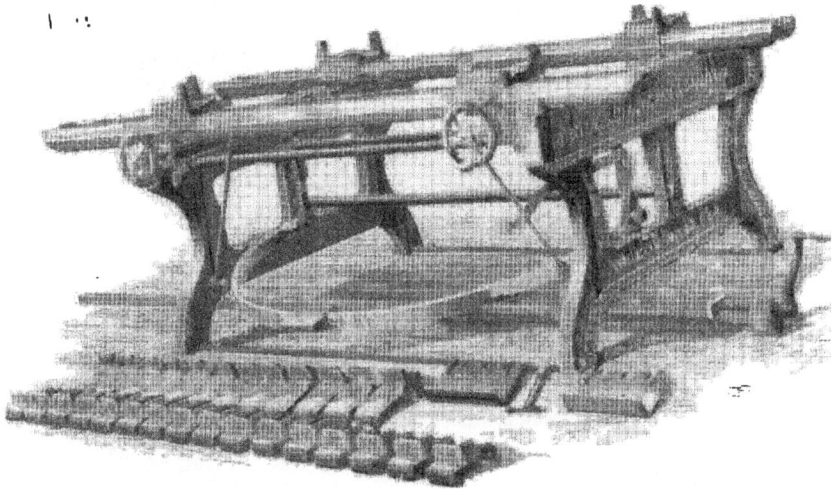

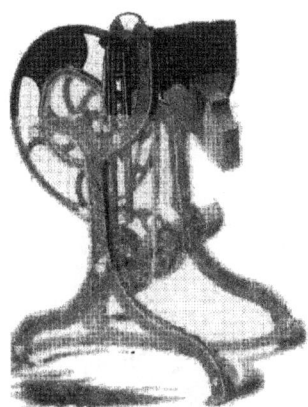

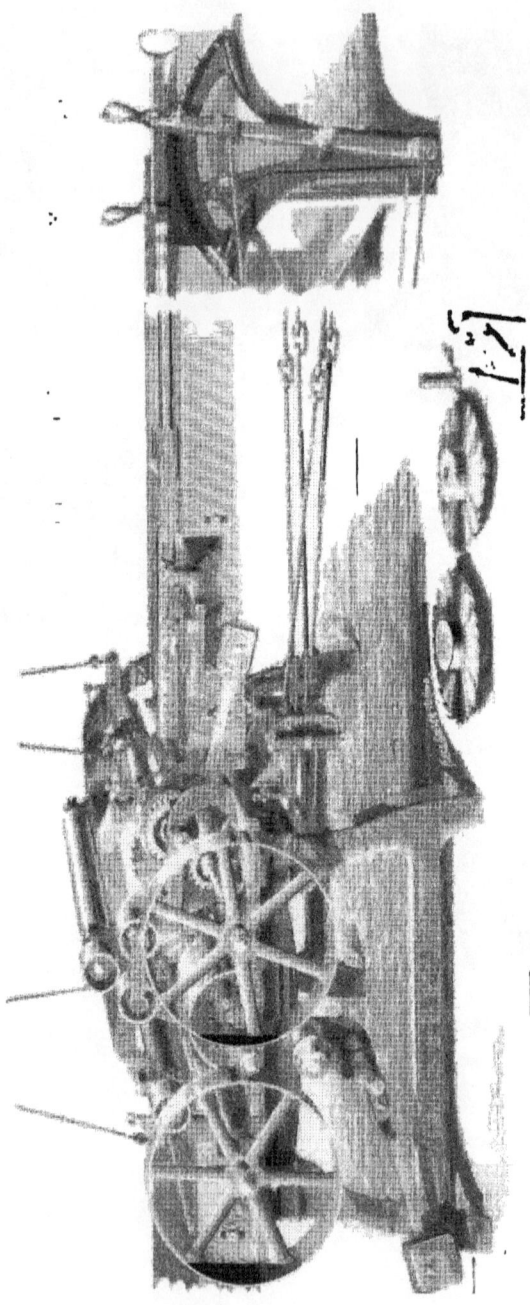

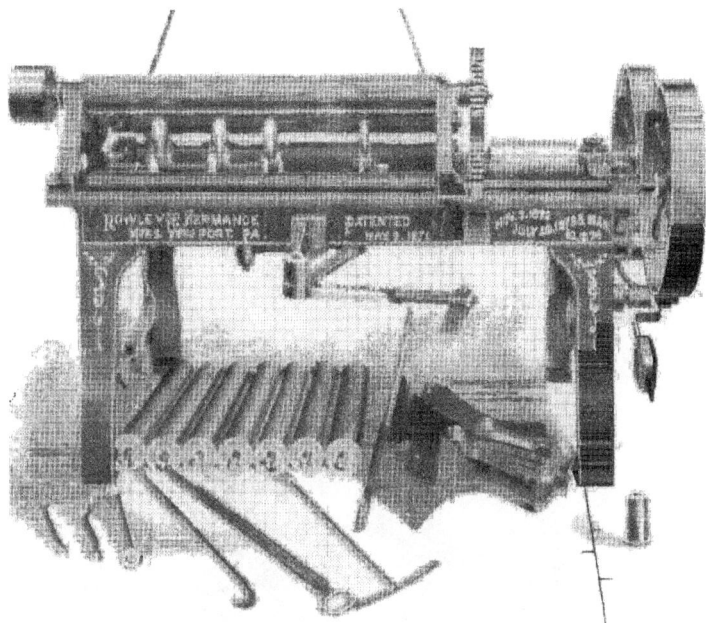

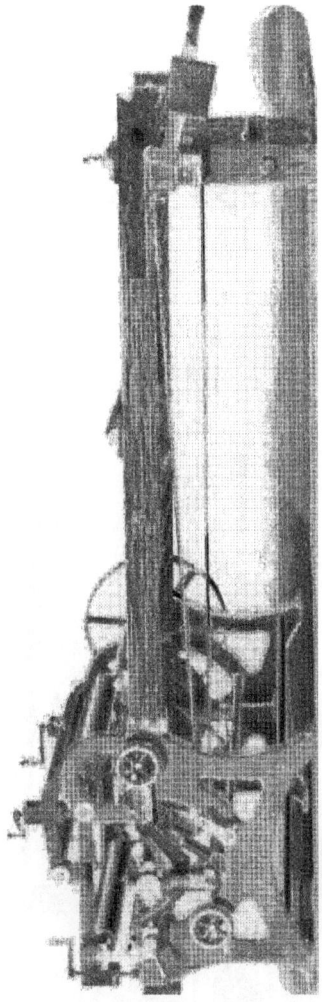

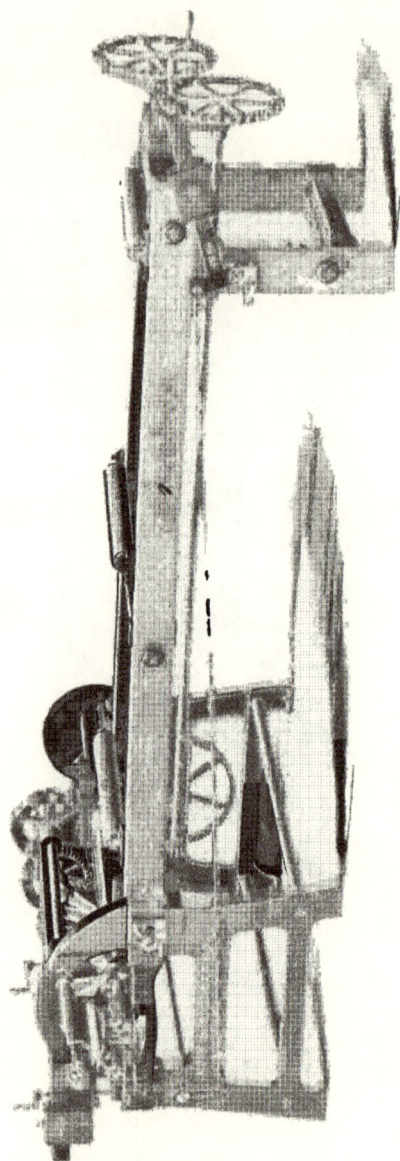

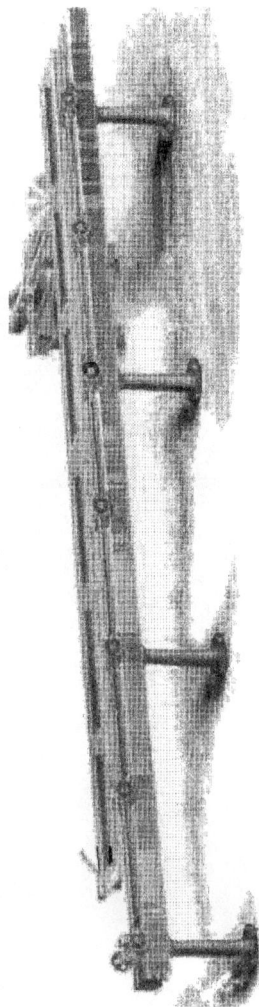

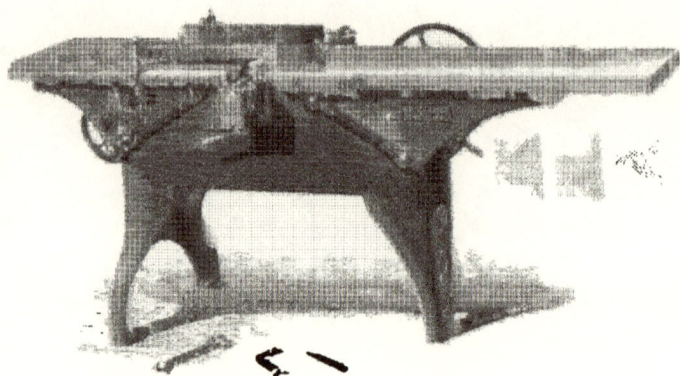

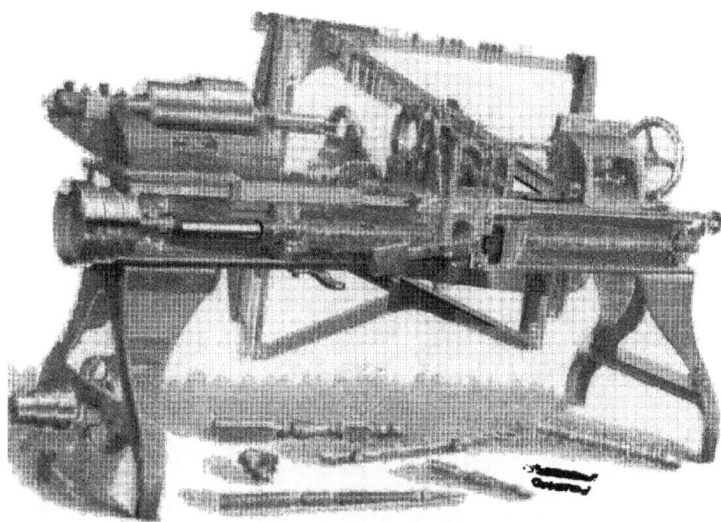

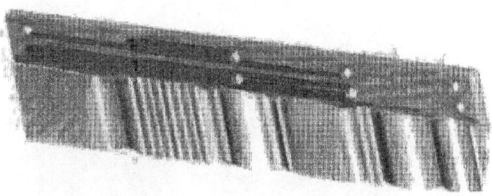

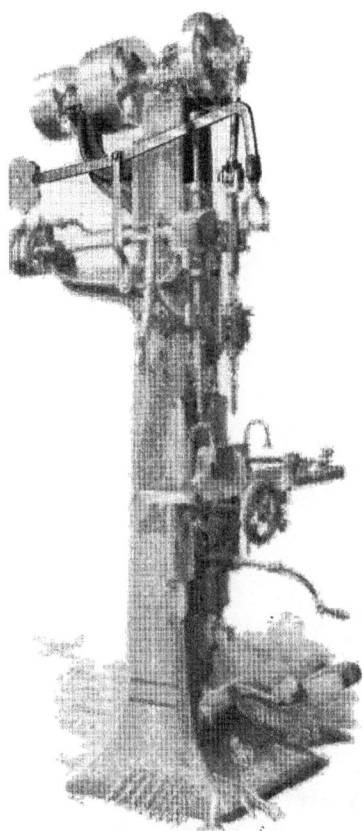

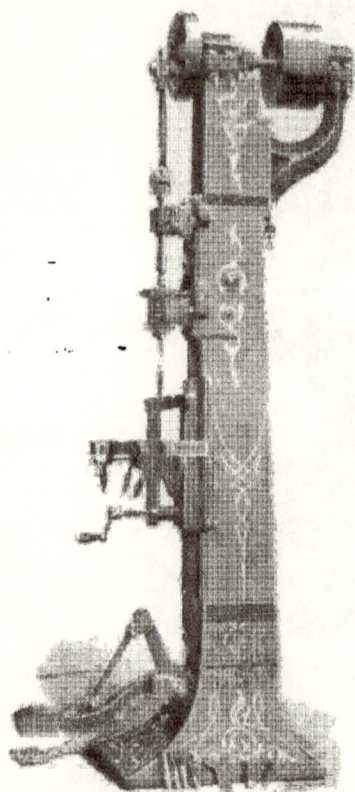

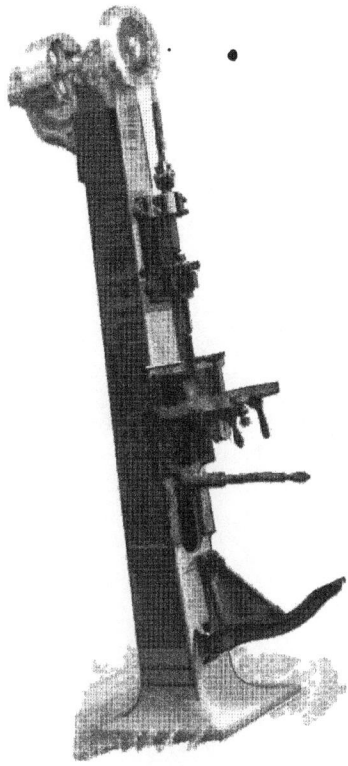

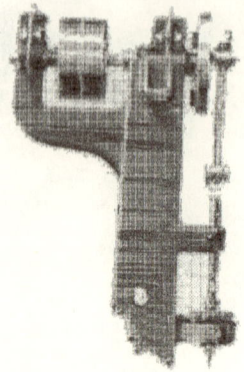

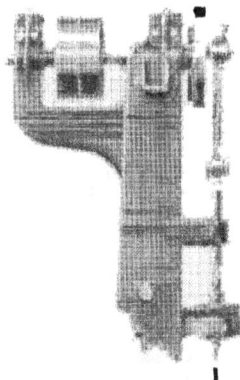

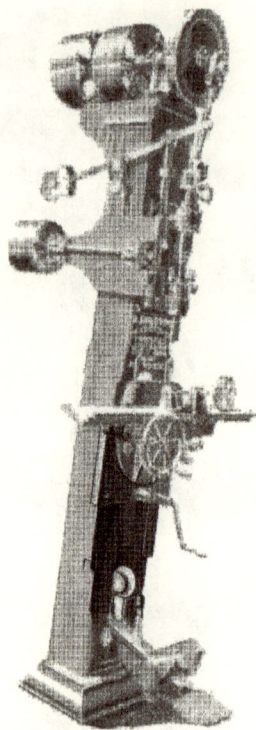

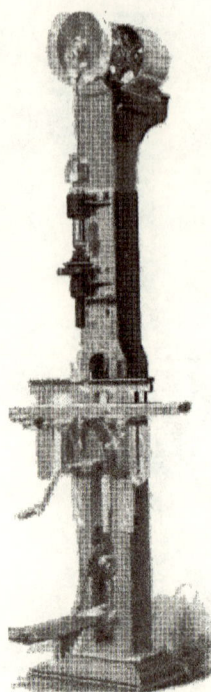

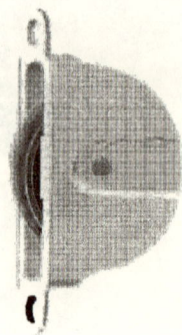

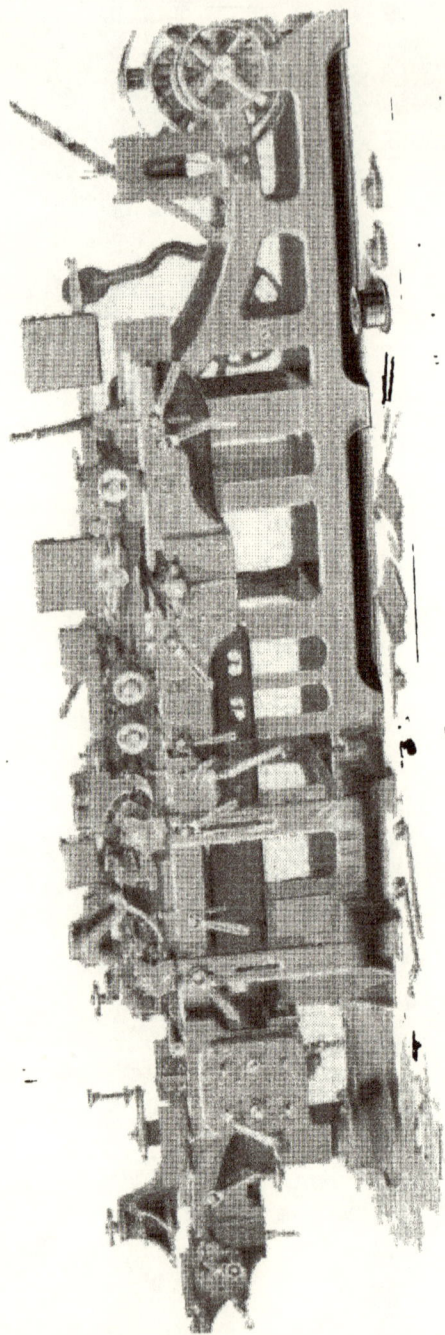

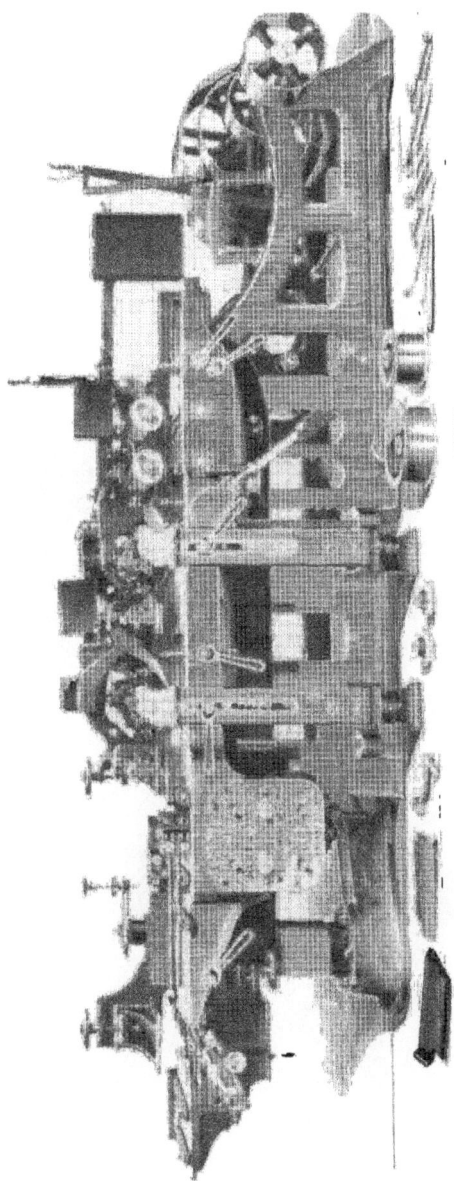

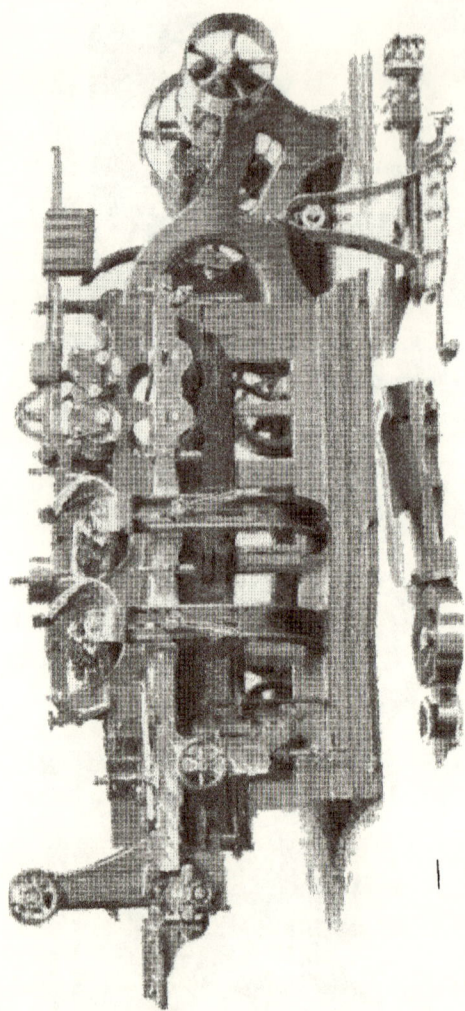

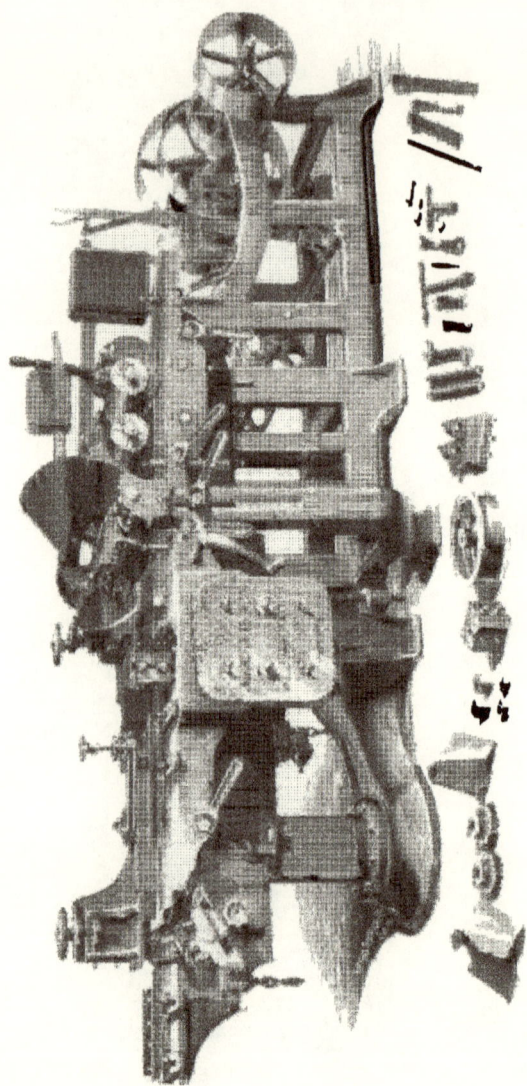

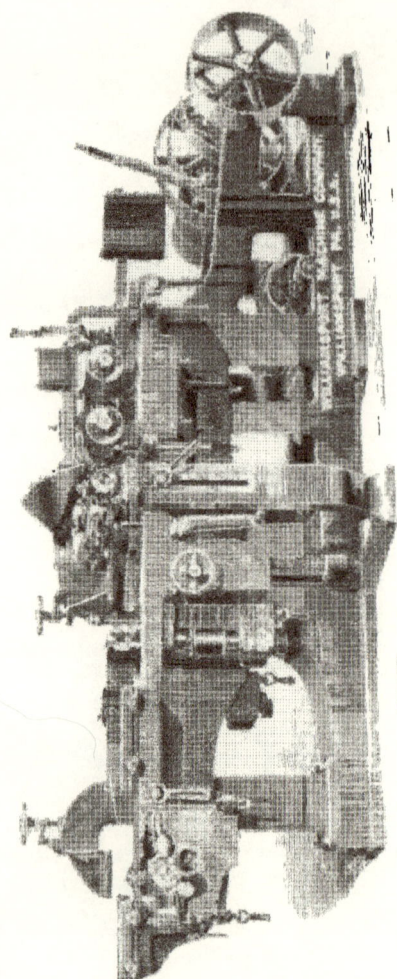

الشكل رقم ٥

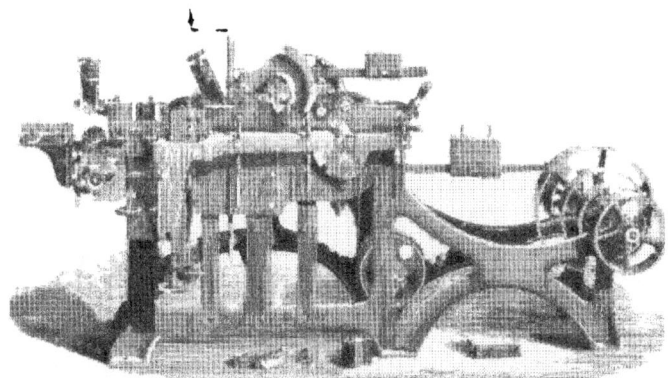

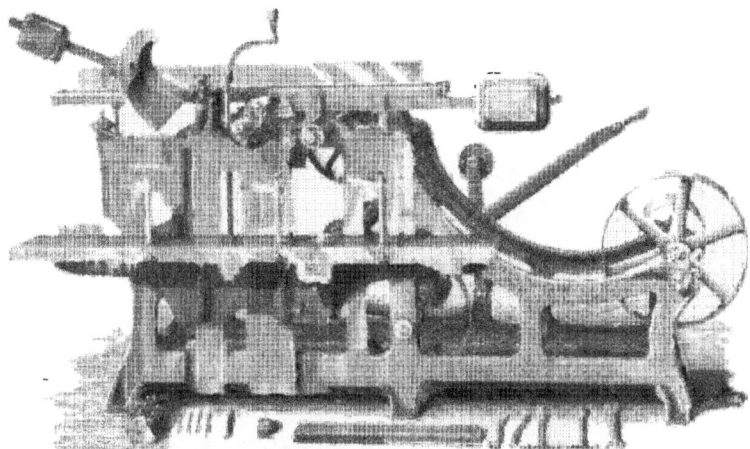

STANDARD TRUST
GLEN COVE MACHINE CO.
LIMITED
BROOKLYN, N.Y.

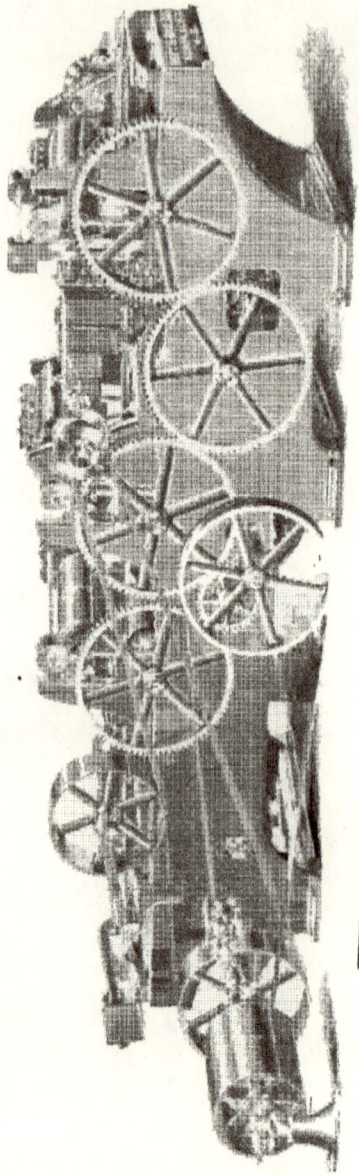

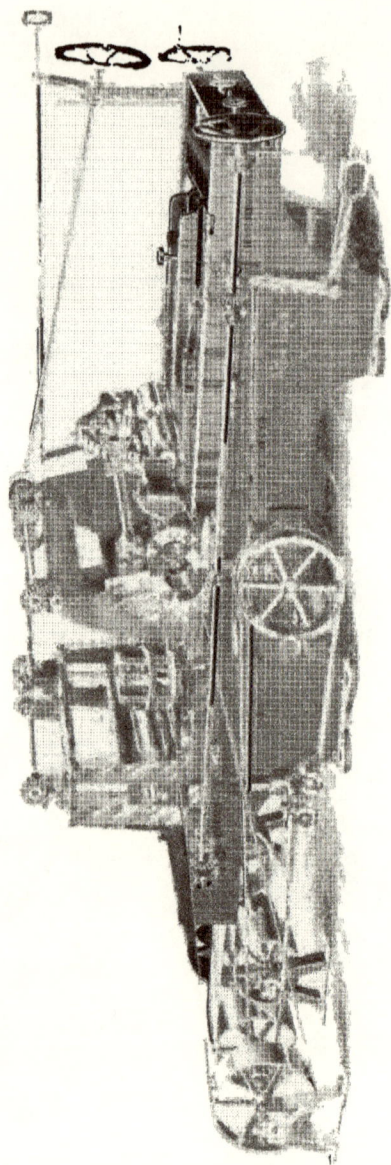

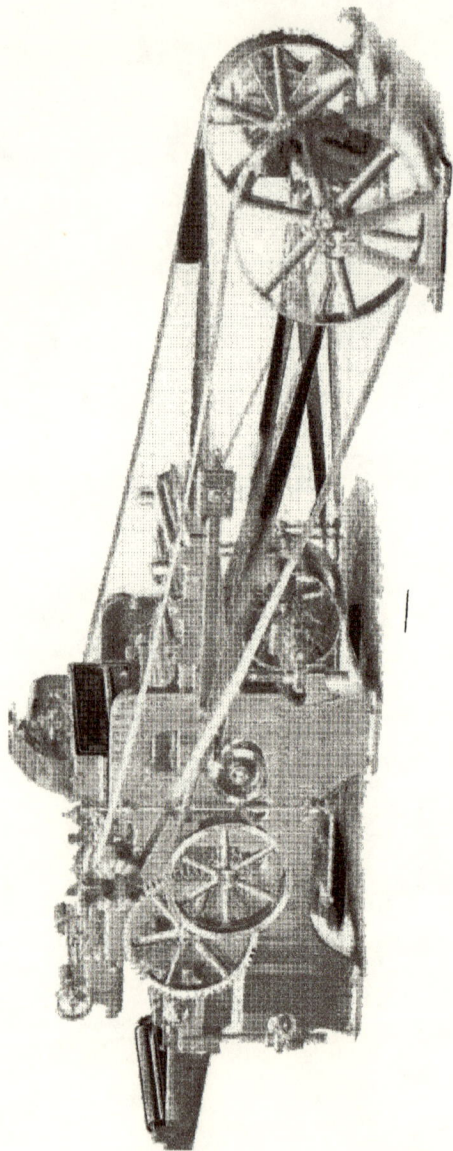

.

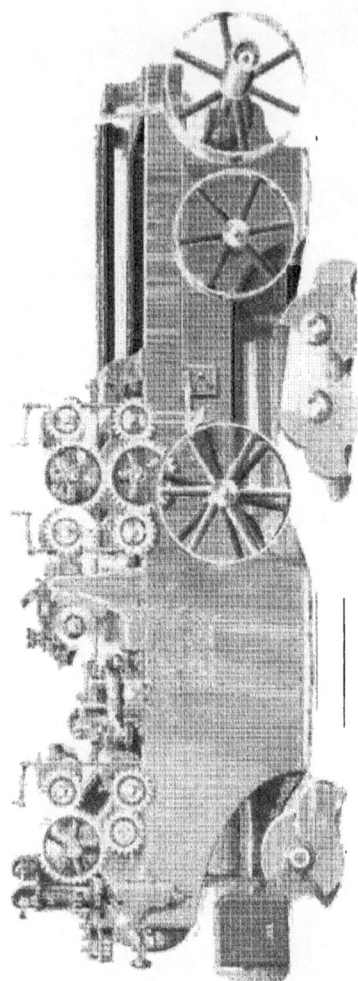

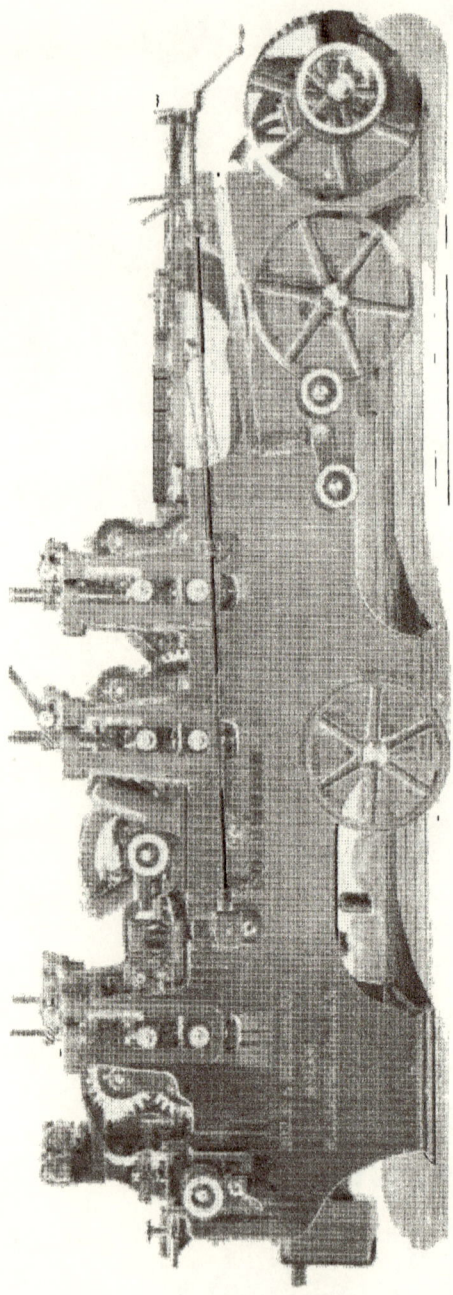

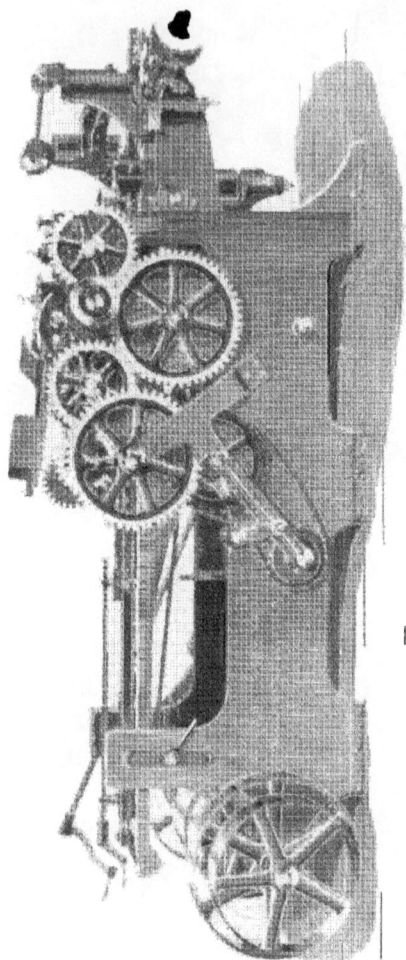

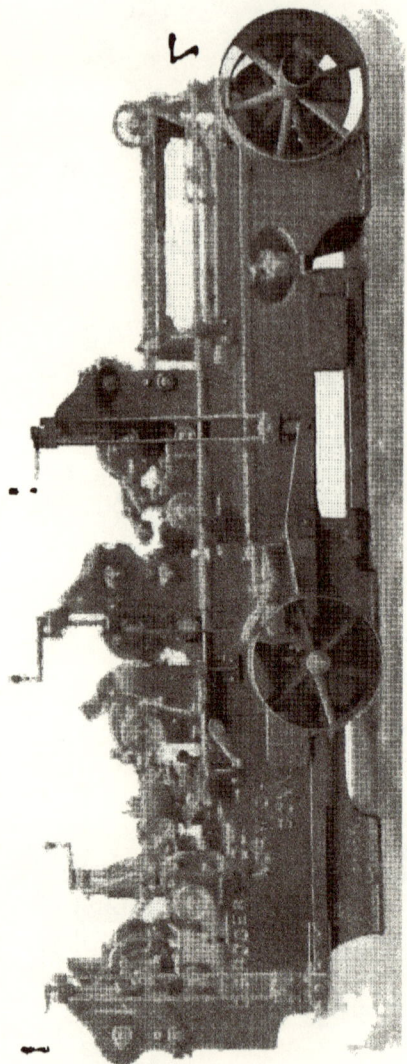

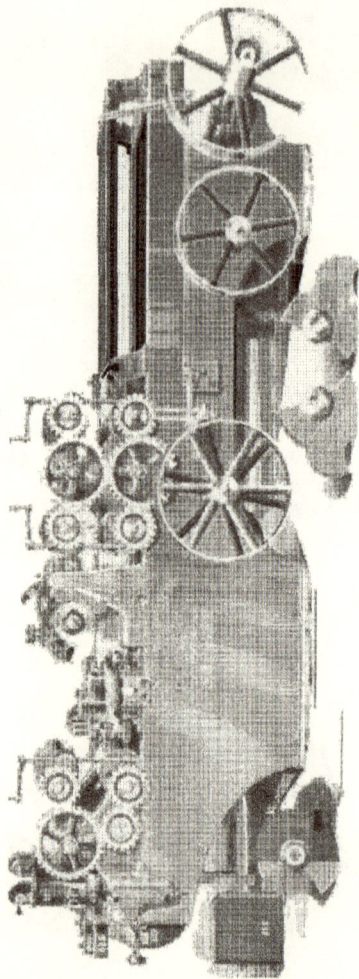

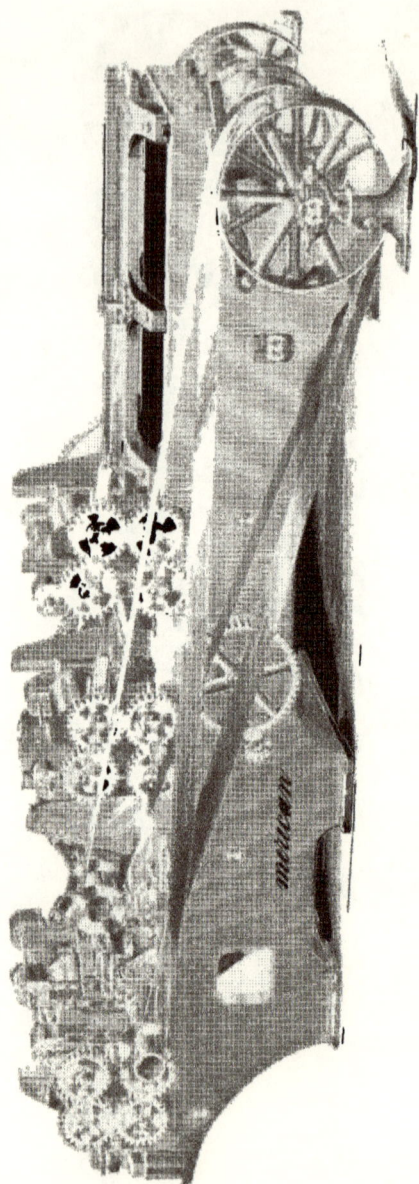

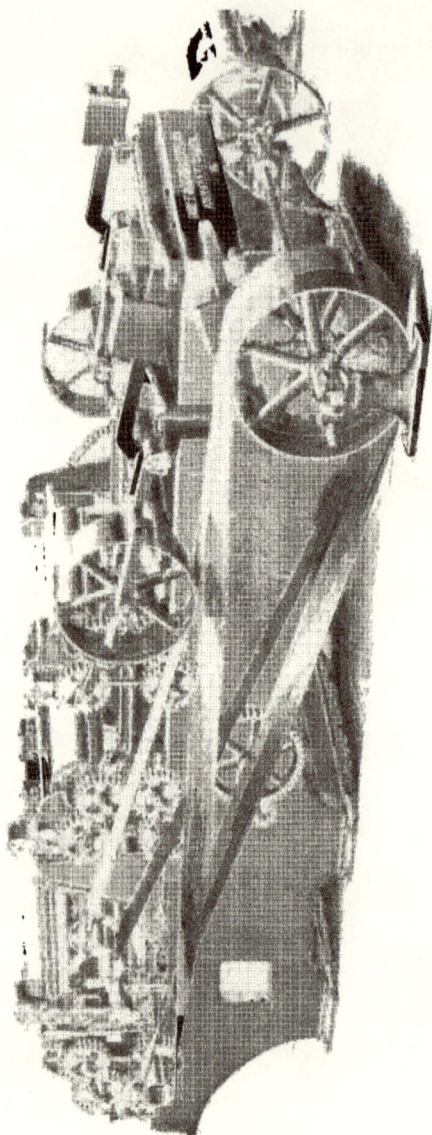

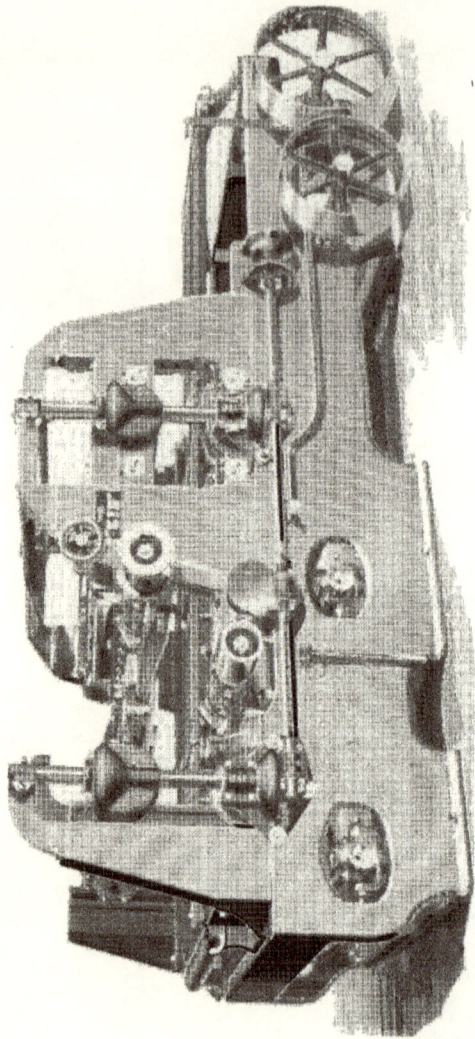

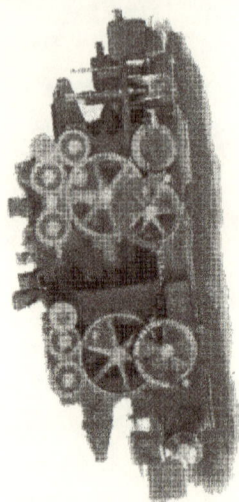

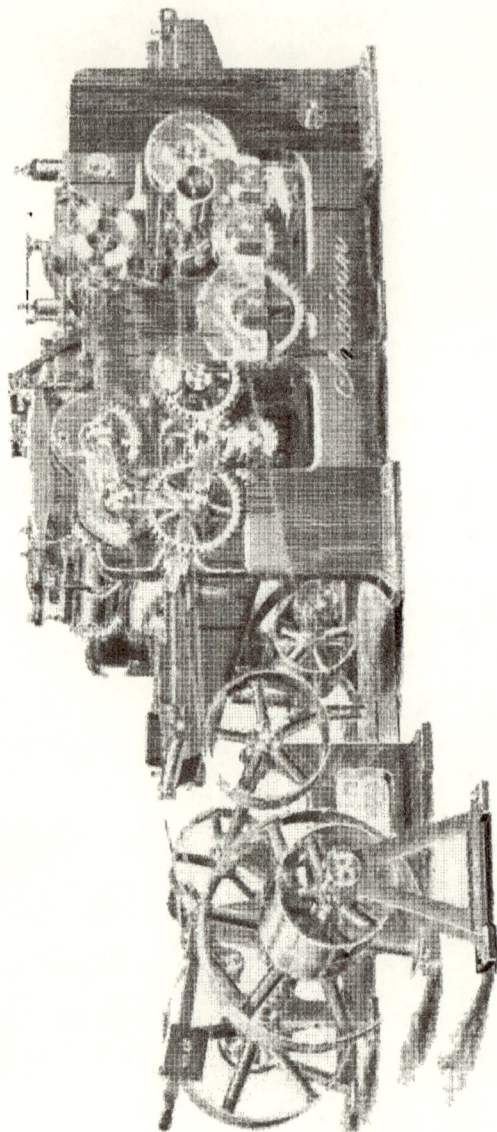

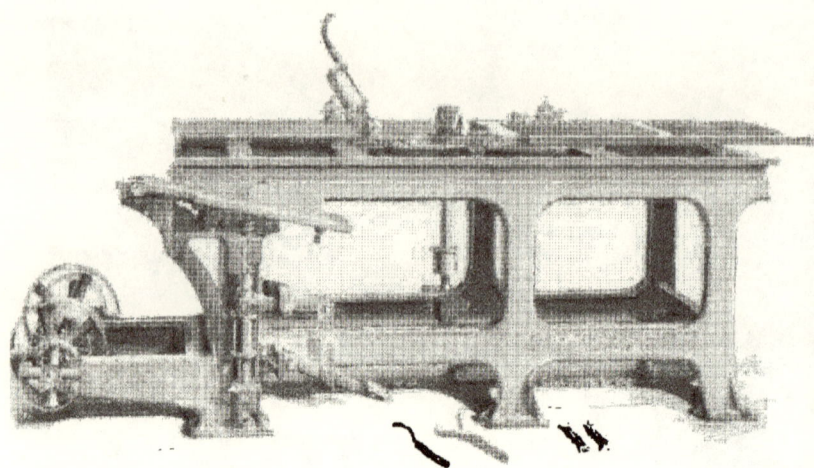

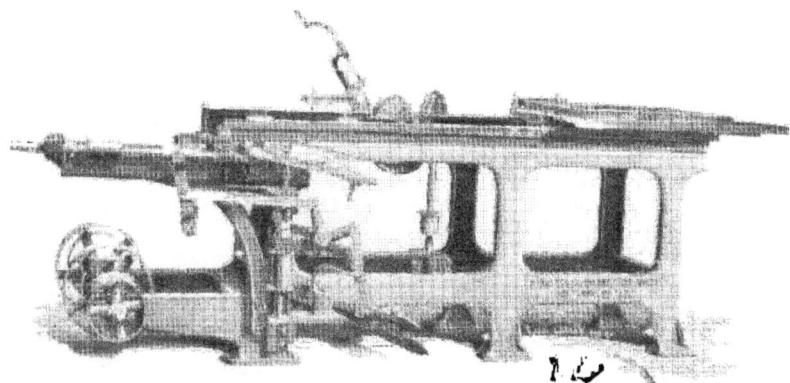

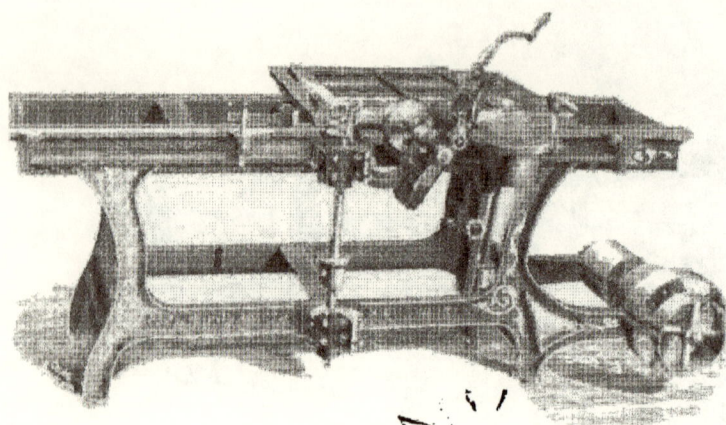

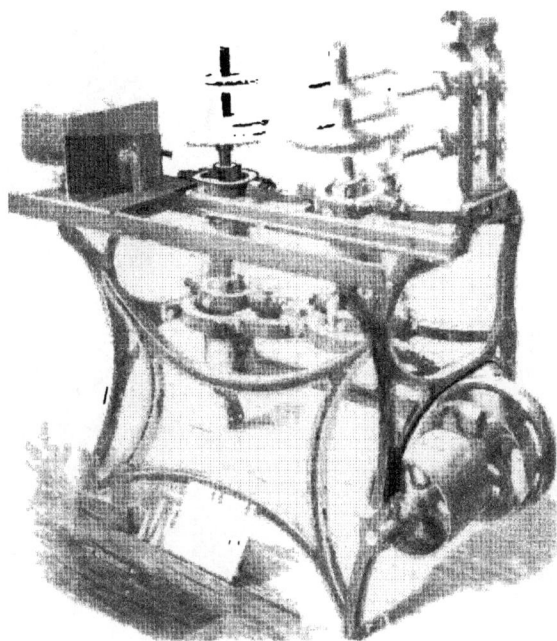

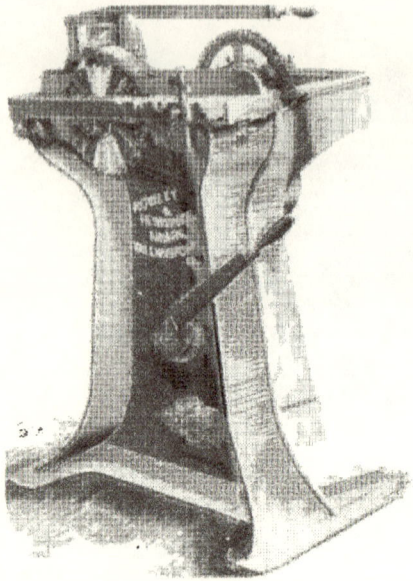

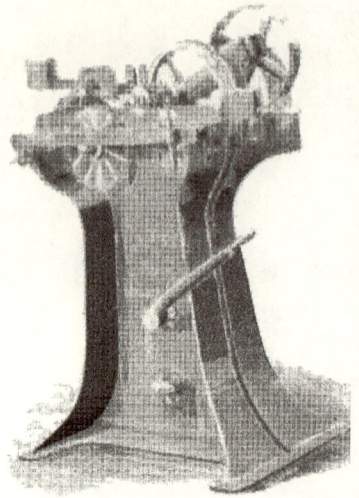

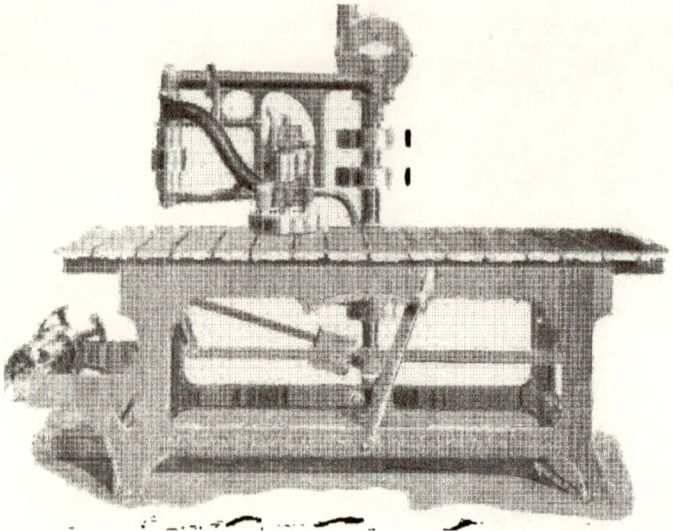

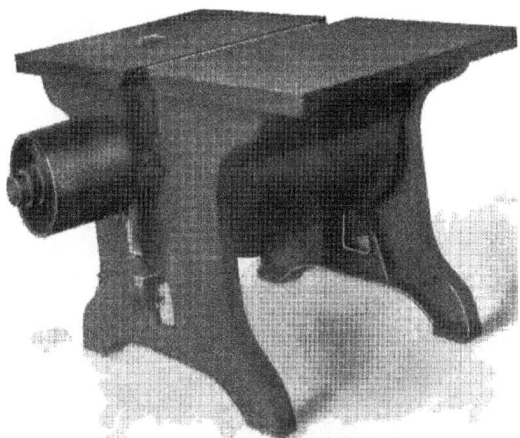

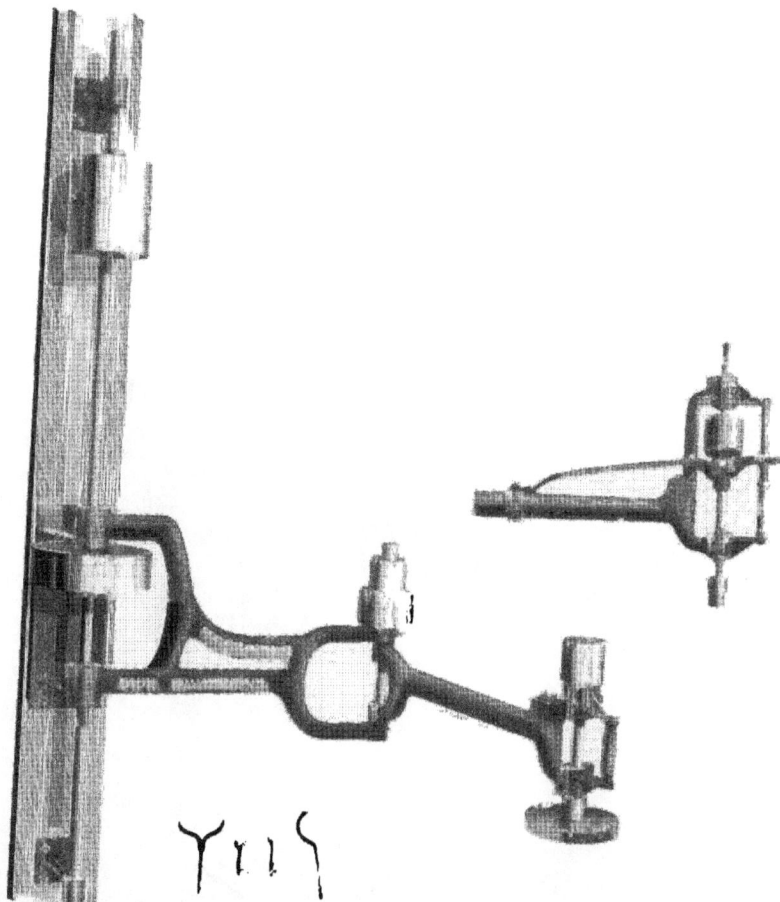

٢٦٤

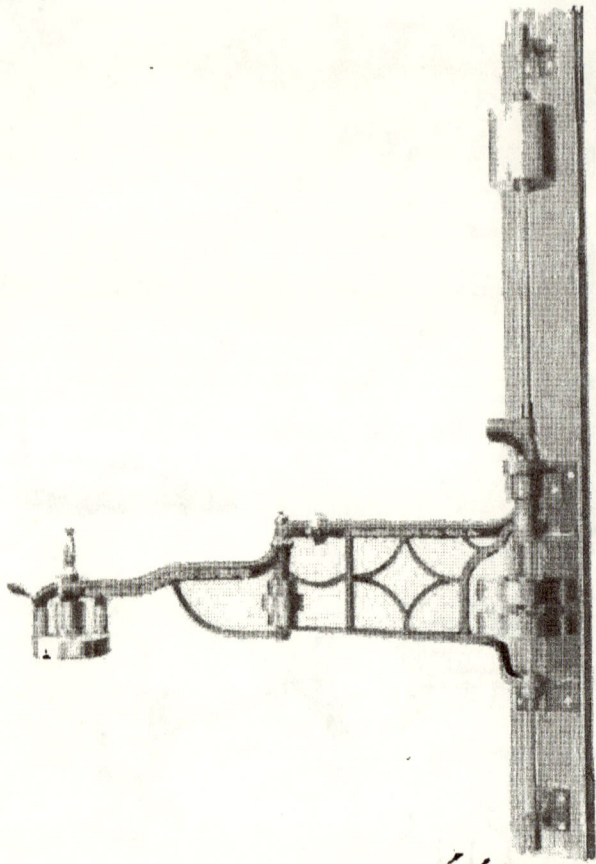

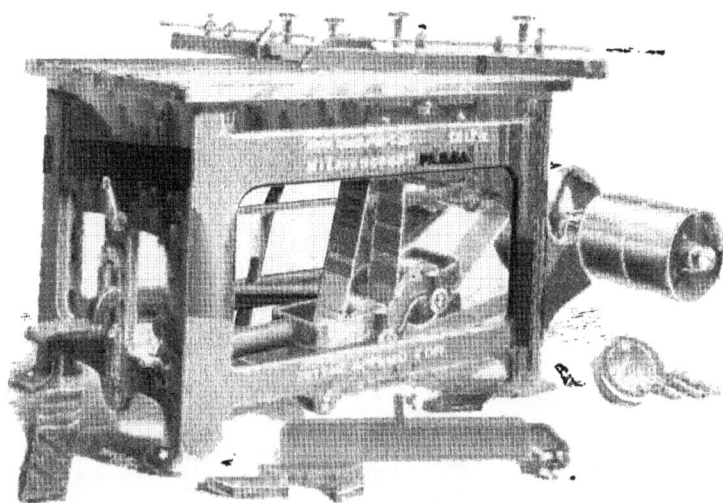

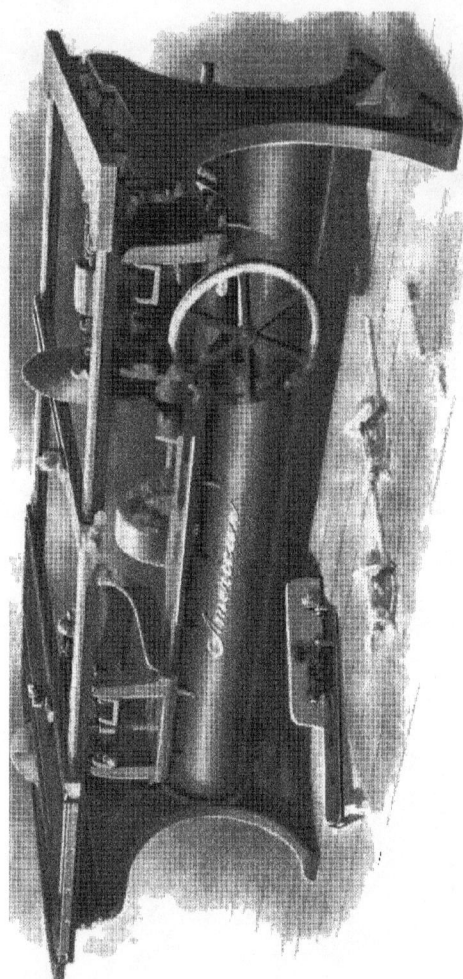

American

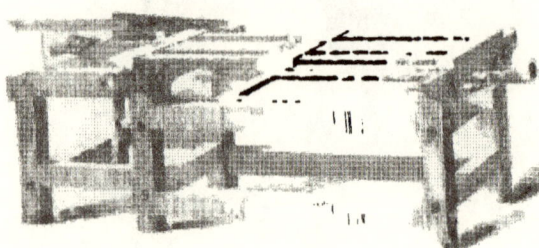

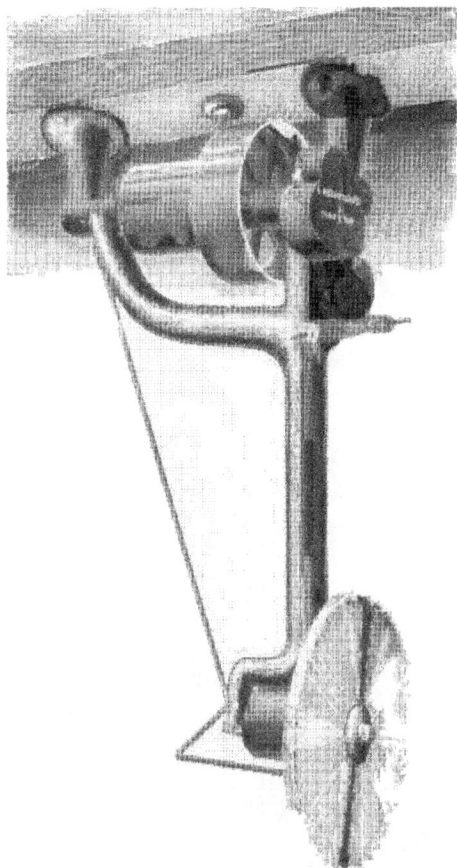

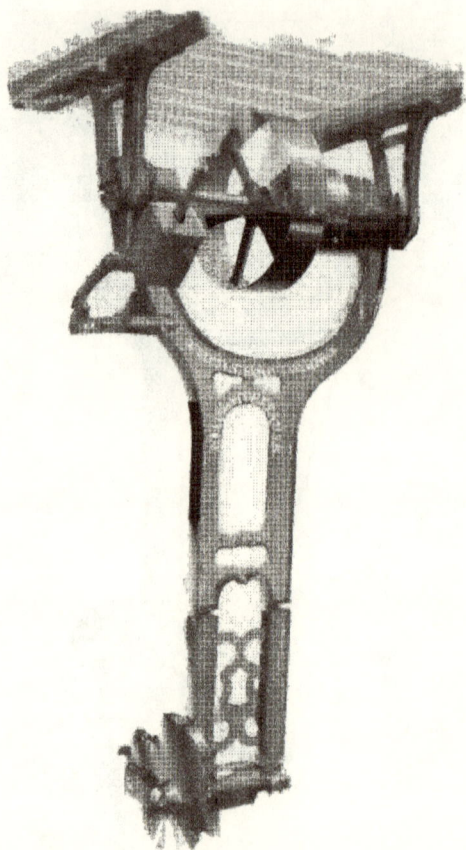

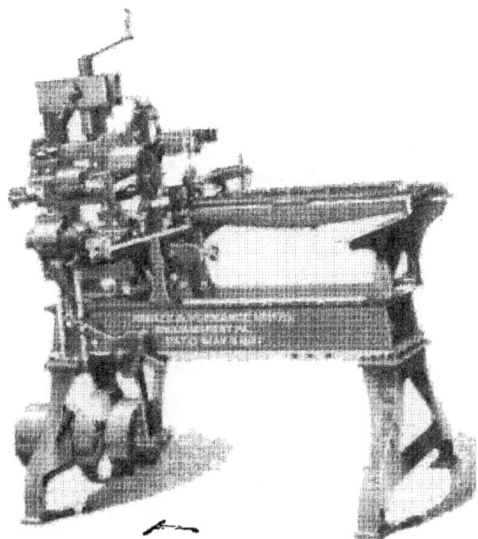

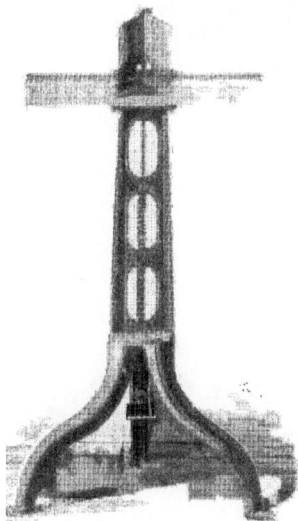

www.ingramcontent.com/pod-product-compliance
Lightning Source LLC
Chambersburg PA
CBHW021402210326
41599CB00011B/977